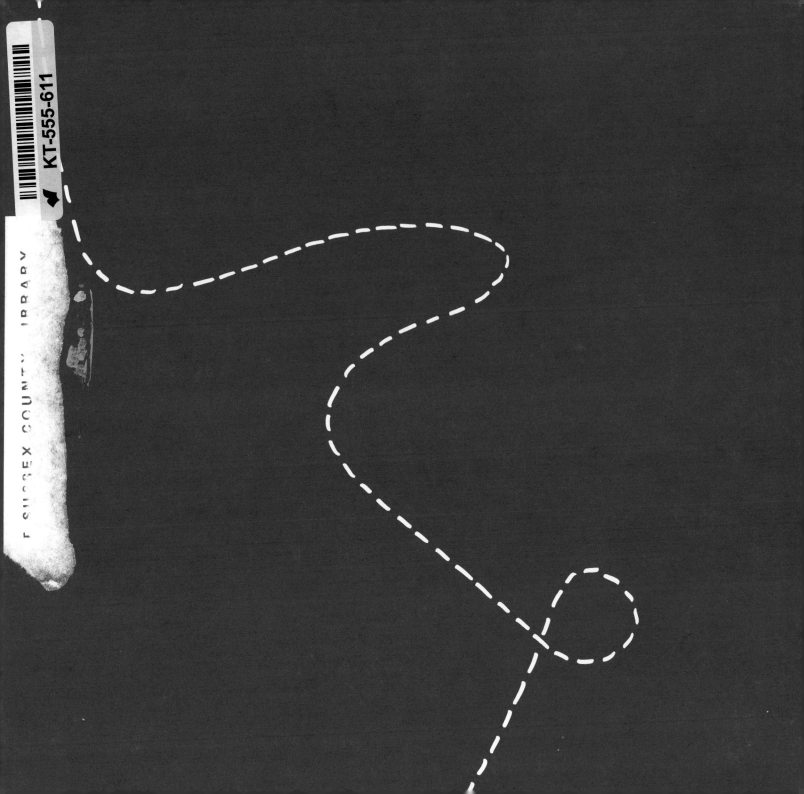

KT-555-611

E SUSSEX COUNTY LIBRARY

LONDON BOROUGH OF WALTHAM FOREST
PUBLIC LIBRARIES
WALTHAMSTOW
47
EAST

BUSY PLACES

Building Site

Carol Watson

With thanks to Wates Built Homes

W
FRANKLIN WATTS
NEW YORK•LONDON•SYDNEY

EAST SUSSEX
V W&S INV. No. 821800 7
SHELF MARK
COPY No
BRN
DATE LOC LOC LOC LOC LOC
COUNTY LIBRARY

It's morning at the building site. Everyone is hard at work doing the many jobs that are needed to build houses.

Gareth is driving a fork-lift truck.
He is using it to lift a heavy load
up to the builders who are working on the roof.

The site manager, Steve Green, is busy checking the architect's plans. These show exactly where and how the houses should be built.

Steve rings up one of his suppliers.
"Is that Mason's Brick Yard?" he says.
"We need another delivery of bricks."

Up on a roof, Brian is putting up scaffolding. The metal tubes make a framework around the building. Brian lays boards over these for the builders to walk on.

Brian turns the bolts tightly to make sure all the scaffolding is safe. Like all the builders on site he wears a safety helmet and special boots to protect his head and toes.

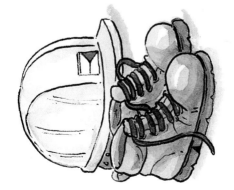

Down on the ground
Jason is using a cement mixer
to make mortar for the bricklayers.

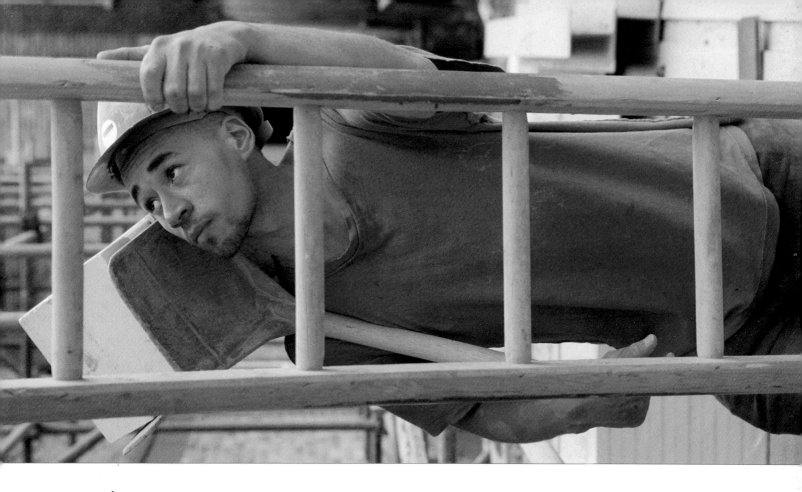

"Can you bring up some more bricks?" calls one of the bricklayers. Jason puts the bricks in a 'hod' to carry them up the ladder to where they are needed.

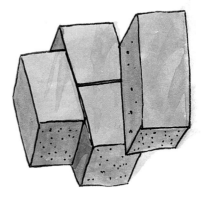

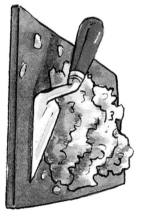

Tony is using red bricks to build the outside walls of a house.

He puts the mortar in place with a trowel. "Can I have another bucket of mortar, Jason?" he asks.

To make sure
that each row
of bricks is
straight, Tony
runs a string
line from one
corner of bricks
to another.
He uses this as
a guide.

12

A carpenter is busy
sawing wood
to make windows.
He has to measure
each piece carefully.

Steve checks how the roofing is going with Bill, the bricklayers' foreman. "We'll soon be finished," he tells Steve.

Meanwhile, on another part of the site a crane driver is at work. He is moving the crane hook to lift some heavy concrete blocks.

The crane driver has a 'banksman'
to show him where to place the load.
"Take it up slowly," signals the 'banksman'.

Some of the houses are almost
finished on the outside.
But inside there are still many jobs to be done.
The builders working indoors
can now take off their safety helmets.

"That looks smooth now," he says to himself.

The plasterer is putting the finishing coat of plaster on the walls.

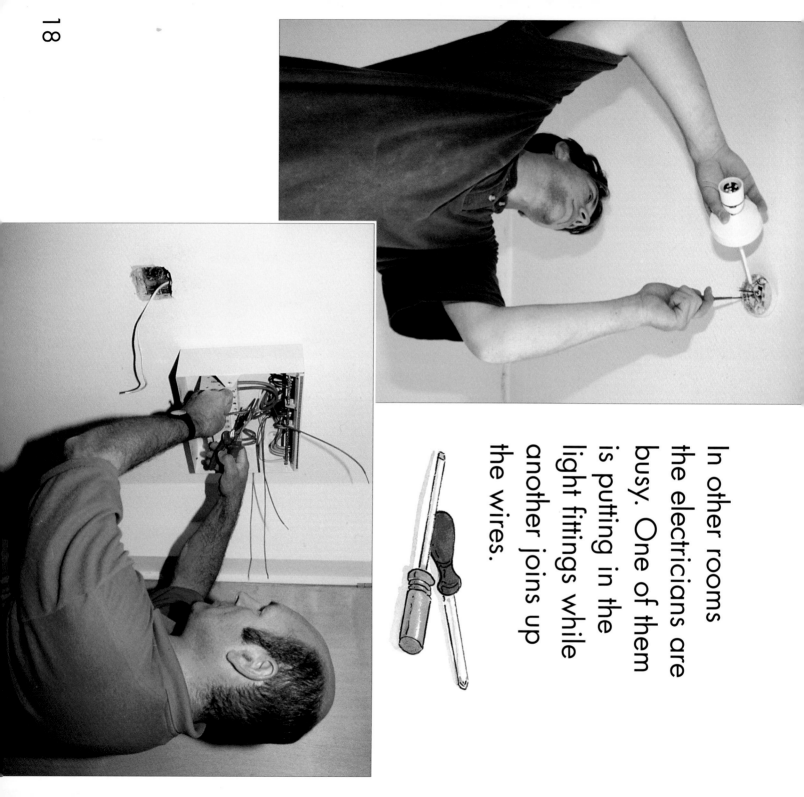

In other rooms
the electricians are
busy. One of them
is putting in the
light fittings while
another joins up
the wires.

The plumbers work on the central heating system.

Meanwhile a carpenter fits the locks into the doors. "That's it," he says. "We'll soon be ready for the decorators."

Safety helmets
must be worn
in this area

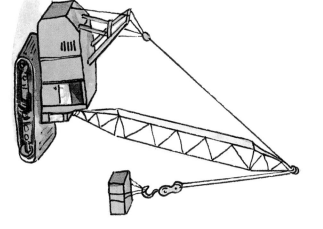

Outside the building
site William and
Hannah are
watching the
machines. "We're
not allowed in
there," says Hannah.
"It's too dangerous."

The children look up at the sign which shows all the things the builders must wear on site to keep safe.

wates
BUILD WITH CARE

SAFETY ON SITE

Use ear protection where appropriate

Foot protection must be worn at all times on this site

Safety helmets must be worn at all times on this site

Wear masks where appropriate

Use hand protection where appropriate

Wear eye protection where appropriate

PARENTS TAKE CARE

William and Hannah go to the sales office to look at the model of the finished site. "I want a big house at the back," William tells his sister. "Would you like a flat at the front?"

Care and safety on a building site

Children are not allowed on building sites because they can be dangerous places. Always **keep out** of all building sites.

These are some of the reasons why:

1. You may fall into holes or trenches.

2. Stacks of bricks may fall on top of you and piles of sand can be dangerous.

3. Because you are small you may not be spotted by the drivers of cranes and trucks.

4. There may be no-one around to help you if you hurt yourself.

5. People who work on building sites have had special training. They wear safety helmets, special boots and other protective clothing. Children do not have these things.

Index

© 1998 Franklin Watts
96 Leonard Street
London
EC2A 4RH

Franklin Watts Australia
14 Mars Road
Lane Cove
NSW 2066

ISBN 0 7496 2994 0

Dewey Decimal Classification Number
690

A CIP catalogue record for this book is
available from the British Library

Printed in Hong Kong

Editor: Samantha Armstrong
Designer: Kirstie Billingham
Photographer: Steve Shott
Illustrations: Richard Morgan

With thanks to Steve Green,
John Winney, Jonathan, William and
Hannah Spencer and all the staff at
Wates Built Homes.